EPAGA HOUSE PUBLISHING COMPANY

Jamantha Williams Watson

COUNTING

IN

NEPALESE

Numbers 1-20

Copyright © 2015 by Jamantha Watson

About the Author

It was a Tuesday afternoon, the class had just come in from recess. The temperature in Buffalo, New York was frigid that day; as snow fell in heavy clumps. It was the perfect time to write an essay. While sitting in Ms. Mumm's English class, Jamantha Watson wrote her first short story. "My Family's Camping Adventures" detailed her family's hiking trip into the Alleghany Mountains. Not only did Watson's English teacher find the story amusing, her classmates did as well. Writing and reading that story peaked an interest in storytelling.

A voracious reader, by eighth grade Watson had devoured nearly all of the children's books and plays in the Buffalo Public Library. She has enjoyed consuming gazillions of books since then. At a young age her parents instilled the importance of both writing and reading and the incredible nourishment they add to an individual's overall development.

One of Jamantha Watson's most inspiring quotes was written by Dr. Seuss. "You have brains in your head. You have feet in your shoes. You can steer yourself any direction you choose.

EK

(eka)

DUI

(dui)

TIN

(tina)

CHAR

(cara)

PAANCH

(pamca)

CHHA

(cha)

SAT

(sata)

AATH

(atha)

NAU

(nau)

DAS

(dasa)

EGHAARA

(eghara)

BARHA

(bahra)

TERHA

(tehra)

CHAUDHA

(caudha)

PANDHRA

(pandhra)

SORHA

(sohra)

SATRA

(satra)

ATHARA

(athara)

UNNAIS

(unnaisa)

BIS

(visa)